The Pet Detective Series

ARE RABBITS THE RIGHT PET FOR YOU

Can You Find out the Facts?

5m Publishing

First published 2015

Copyright © Emma Milne 2015

Published by

5M Publishing Ltd,
Benchmark House, 8 Smithy Wood Drive, Sheffield, S35 1QN, UK

Tel: +44 (0) 1234 81 81 80
www.5mpublishing.com

A Catalogue record for this book is available from the British Library

ISBN 978-1-910455-05-0

Book layout by Mark Paterson
Printed by CPI Group (UK)
Photos by Dreamstime
Cover photos RWAF

Pet Detectives: Rabbits

Can *You* Find Out the Facts?

By

Emma Milne

BVSc MRCVS

For Alice and Charlotte.

Sorry I'm a better vet than I am a mum!
I love you all the way to the moon. And back.

Contents

Acknowledgements

This book would not have been possible without such fantastic support, photos and help from lots of people. Massive thanks to Richard and Rae at the Rabbit Welfare Association and Fund (RWAF). Your support, endorsement, advice, proofreading and photos are hugely appreciated. Thank you to the Animal Welfare Foundation. Having endorsement from such a respected charity is a great privilege. Thank you to Emily Coultish for bringing my cartoon ideas to life and for putting up with my pedantry! Finally, a huge thank you to the PDSA, Marit Emilie Buseth, author of *Rabbit Behaviour, Health and Care*, Sean McCormack MVB and Dr Estella Böhmer for granting the use of their wonderful photographs. I salute you all!

The author supports the promotion of animal welfare issues and welfare related activities of many organisations globally including IFAW (International Fund for Animal Welfare), HSUS (Humane Society of the United States), World Animal Protection and the Pet Education Alliance to name but a few.

Chapter 1

Humans and animals have shared this lovely little planet of ours for thousands of years. Over that time we've used and needed animals in all sorts of different ways. First of all we needed their bodies and we used to use all the bits we could. We ate their meat and some of the other bits too, like kidneys and livers and hearts. We got rich nutrients out of their bones, from the bit called the marrow. We used their stomachs to carry water in or to cook other bits in. Animal fat was used to make oils and candles and fuel lamps and torches. Bones and horns could be made into tools and cups and we used animal skins and fur to keep us warm and dry. Even things like a bison's tail could be made into anything from a fly swat to an ornament.

"Hey! I'm still using that!"

These days we still eat plenty of animals but we've found substitutes for some of the other bits. Over those thousands of years we started to realise that animals could be useful for other things. Cats were very good at catching mice and rats and other pests which ate our crops, so having cats around started to look like a good thing. They could do jobs for us that we found hard. Wild dogs started creeping closer to human camps to get some warmth from our fires and steal scraps of our food. But in return humans got some protection from their natural guarding instincts and we learned that if we joined them in hunting we all made a pretty good team. Horses, donkeys and camels could be tamed and could carry big loads and cover distances which man couldn't even attempt. These animals are still used all over the world today.

As more time has gone by we've stopped needing animals so much to do things for us. We have cars and tractors, we have mouse traps and food containers that can't be chewed through, we have houses with alarms and strong doors and big locks and we have farmed animals to eat so we no longer need to hunt. But the simple truth is that we found out that animals are wonderful creatures. In the time we were getting to know each other, humans started to fall in love with just having animals around.

A dog putting its head on your lap to have his ears stroked gets some lovely affection but the human who's feeling those velvety ears and looking into those lovely brown eyes gets a lot back too. The cat stretched out on a warm rug by the fire has definitely 'landed on its feet' but the doting owner — smiling in the doorway, just watching the cat's tummy rise and fall with its breathing — feels happy without even realising it.

Pets make us happy. They make us feel calm and loved and wanted. Pets don't judge us or hold a grudge for days like your best friend did when you spoke to the new girl at school.

They are always there and they stick with us through thick and thin. In fact, sometimes they seem a lot nicer to be around than some humans!

The Serious Bit

The last point is the bit that is so important to remember; humans don't always do the right thing for their pets. Pets don't get to decide who buys them or how they are cared for. They have to live wherever you put them and they can only have the food you give them because they can't get to food themselves. When they go to sleep, how comfortable their bed is will be totally up to you. What you have to realise is that if you want pets, those animals are *completely* dependent on you and your family to keep them happy, healthy and safe.

It sounds like an easy thing, doesn't it? Buy a cage, or a cat bed or a dog's squeaky toy, go to the pet shop and buy a bag of food and your pet will have everything it needs. WRONG! Thousands – if not millions – of pets have been kept this way, and of course with food and water most animals can survive for years; but that is just not enough. A great life isn't about coping or managing or *surviving*. It's about being HAPPY! If you were locked in your bedroom with no toys or books or friends or even your pesky sister you'd *manage*. As long as your mum gave you bread and stew twice a day and the odd bit of fruit you'd probably live for years. But would you be happy? It doesn't sound likely, does it? You'd be bored out of your mind, lonely, miserable and longing for someone to play with, even if it was just that nose-picking sister or brother that you usually avoid like a fresh dog poo and make cry in front of your friends.

Having a pet, any pet, is a serious business. It's a bit like getting married or having a tattoo; you definitely shouldn't rush into it! You need to think carefully about lots of things. What sort of house or flat do you live in? How big is your garden if you have one? What other animals, if any, have you already got? How much money do your parents earn because I can tell you there is no such thing as a cheap pet. How much spare time do you *actually* have? Are you an active family or a lazy one? All these questions have to be asked, and they have to be answered very *honestly*. And of course you need to ask yourself what sort of animal you actually want.

I've been a bit sneaky there because actually you should *never* ask yourself what sort of animal you want. You should think about what sort of animal you can look after properly. There's a very famous song from a long time ago called *You Can't Always Get What You Want*, and I'm sure your mum or dad will have said it to you hundreds of times. You probably rolled your eyes, walked off in a huff, slammed a door and shouted 'THAT'S NOT FAIR!!' But I hate to say that your mum and dad are right – and it's especially true when it comes to keeping pets. Lots of animals get abandoned or given away because people don't ask themselves the right questions, don't find out the facts and then, most importantly, don't answer the questions honestly.

Let's be honest, you lot are masters at pestering. For as long as children, parents and pets have been around, children have pestered, parents have caved in and pets have been bought on an impulse! This means without thinking and without knowing what the animal actually needs in order to be happy. This usually means a very miserable pet.

But we're about to change all that aren't we? Because now I've got the dream team on my side. You chose to find out the facts about these animals so you *could* make the right choice. And I am very proud of you for that and I am very happy. So thank you.

The EVEN MORE serious bit!

So, you are thinking it would be nice to have a pet. You're certain you are going to love it, care for it, keep it happy and of course, NEVER get bored with looking after it and expect your mum and dad to do it. But what you need to know is that not only is that the right thing to do — it is also now the law. Sounds serious, doesn't it? As I said, it is a serious business. In the United Kingdom in 2006 a new law was made called the Animal Welfare Act. This law says that anyone over the age of 16 who is looking after an animal has a 'duty of care' to provide for all the needs of the animal. Now, laws are always written by people who use ridiculously long words and sentences that no-one else really understands — but this law is very important to understand. A duty of care means it is the owner's responsibility to care for the animal properly and the law means that if the owner doesn't they could get their pet taken away and even, in rare cases, end up going to prison!

Aha, you may think, I am not 16 — so I'm fine. You'd be right, but the duty of care then falls to your mum or dad or whoever looks after you and the pet. So if you would like some rabbits, not only do you need to know all about them, you need to make sure the adults in the house do too. And you need to make sure they know about the law because they might not know what they are letting themselves in for!

Well, that's quite enough of all the boring serious stuff — let's learn some things about animals! The easiest way to find out about animals is to know about the five welfare needs. These apply to all pets, and in fact all animals, so they are good things to squeeze into that brilliant brain of yours so you can always remember them whenever you think about animals.

The Need for Fresh Water and the Right Food

This is a very obvious thing to say, but you'd be surprised how many animals get given the wrong food. In fact, there was a queen a very long time ago who wanted a zebra. I said 'wanted', didn't I? She definitely didn't ask herself the right questions or find out the facts, because when someone caught her one from the wild, she fed it steaks and tobacco!

Animals have evolved over a very long time to eat certain things and if they are fed the wrong foods they can get very ill, very fat, or miss vitamins and minerals they might need more than other animals. The right food in the right amounts is essential.

The Need for the Right Environment

This is a fancy way of saying 'where the animal lives'. It could be a hutch, a cage, a house, bedding, shelter, a stable or a range of other things, depending on the pet. It's very important that these spaces are big enough, are clean, are safe and secure, and the animals have freedom to move around.

The Need to Be With or Without Other Animals

Some animals live in groups and love to have company. Some animals are not very sociable at all, like me in the mornings! It's very important to know which your pet prefers. If you get it wrong you could have serious fighting and injuries or just a very lonely and miserable pet.

The Need to Express Normal Behaviour

Knowing what animals like to do is really important. As I said before, lots of animals will survive on food and water, but happiness or 'mental wellbeing' is just as important as being healthy or having 'physical wellbeing'. You've probably never thought about your own behavioural needs but imagine how you would feel if you were never allowed to go to the park or play or run or see your friends. You would soon be quite unhappy. Often you find that happy pets stay healthier, just like us.

The Need to Be Protected from Pain, Injury and Disease

Animals can get ill, just like us, and it will be up to you and your family to keep your pet healthy as well as happy. Just like you have vaccinations, they are very important for some animals too, to stop them getting ill and even dying. Animals, just like lots of children, also get worms, lice, mites and other parasites. You will need to find out how to treat or prevent these and look out for signs of them.

You need to check your animals over at least once a day to make sure there are no signs of problems and take them to a vet as soon as you think something is wrong. Vet costs are not cheap. You might also have a pet you can get health insurance for, which is always a good idea.

So now you know the basic needs of all animals, it's time to concentrate on rabbits! Rabbits haven't actually been kept as pets for very long and still have lots of wild instincts and needs. The best way to learn what will keep your pet rabbits happy and healthy is to find out what rabbits in the wild are like. How do they live? What do they eat? Do they like to have others around, and what makes them scared or nervous? In other words, what keeps them HAPPY? Shall we begin?

Chapter 2

RABBITS IN THE WILD

If there's one thing you can definitely say about rabbits, it's this: there are millions of them! Rabbits are definitely a very successful species. This means they have managed to find places to live and reproduce all over the world and in all sorts of climates. You can find rabbits as far north as the Arctic Circle, all the way over in North and South America, in central Africa and Japan, and as far south as the islands around Australia and New Zealand. In fact one of the few places you won't find them is in the Antarctic.

RABBIT FACT:

A group of beautiful Arctic hares.

The polar rabbit isn't actually a rabbit at all. It's another name for the Arctic hare.

Lots of countries, like Britain and Australia, never had rabbits until they were taken there by people. Rabbits used to be a good source of food for humans and even nowadays in lots of countries people still eat rabbits. Because they were a good source of food and quite easy to keep, rabbits ended up being taken to new countries by explorers. The Romans gave them to Britain over 2000 years ago and then Britain (very kindly, you might think) gave them to Australia when we went exploring over there. It might sound like a nice present but in many countries rabbits have become a real pest. They have so many babies so quickly that the number of rabbits explodes, and that means there are a lot of hungry mouths to feed. They can eat huge amounts of human crops and cost farmers an equally huge amount of money.

Now nature is very clever and keeps animal numbers under control really well. It usually only goes wrong when humans interfere and don't think things through! Over the millions of years that there has been life on our beautiful planet, plants and animals have grown and evolved together in a delicate balance, all linked together by the food they eat. Some animals eat just plants. Some, like humans and bears, eat plants and animals. Still others eat just other animals. These relationships are called food chains. For rabbits one food chain might go like this:

The rabbit eats the grass and the fox eats the rabbit. In this food chain the fox is at the top of the chain because nothing eats the fox. So you can see that having foxes around to eat rabbits is a way that nature controls the numbers of rabbits. The same thing applies to every living thing on the planet. You might be wondering how the numbers of the animals at the top of the chains are controlled if nothing is eating them, and it's a very good question. Quite simply, they are controlled by whether or not there is enough food for them to eat. If there isn't, they die. This all might seem a bit sad and a bit cruel when you think about it but this is how life has evolved and it is absolutely essential for keeping the amazing balance of nature.

In food chains, some animals are what we call prey animals and some are predators. Prey animals are ones which are eaten and predators are the ones which eat the other animals. In the case of our simple food chain the rabbit is the prey because it is eaten by the fox. The fox is the predator because it eats the rabbit. Of course, lots of animals are both predator and prey if you think about it. For example, tuna are huge fish which live in the oceans and eat smaller fish so they are predators. But then a human comes along, catches the tuna and eats it in a big sandwich with some mayonnaise and sweetcorn. The tuna is the prey and the human is the predator.

By the way, if you tried to fit the biggest tuna fish in a sandwich it would be roughly the same size *and weight* as a 1970s mini!

So, in countries like Britain and Australia where rabbits didn't live naturally there weren't enough predators like foxes, weasels and birds of prey to keep the numbers down. Without this natural control they just kept multiplying, invading crops and making a nuisance of themselves. But really, we only have our interfering selves to blame!

RABBITS IN THE WILD

Now we know that rabbits are prey animals and this is really important because it explains lots about the way they live, how they behave and what makes them feel happy, safe and secure. Here is a very long word that's difficult to say; anthropomorphism. Mary Poppins would probably manage it with no trouble at all and be able to make it into a song! It's an important word to know because lots of people who work with animals say you shouldn't do it. It means thinking that animals are just like humans. In general you shouldn't do that because animals are very different to humans and if you assume they like or dislike the same things as you, you could get it very wrong and end up neglecting your pets. Having said that, when you're finding out real animal facts (as you are doing by reading this book) I think it's good sometimes to think about how you might feel in a certain similar situation. By finding out how animals choose to live you can compare it to things in your life and it helps you to empathise with your pets. You'll see what I mean as we go along.

That scared feeling and all the things that go with it like the racing heart are because of something your body makes called adrenaline. It's an amazing hormone, released in an instant whenever an animal (or a human) feels threatened. It's often called the 'fight or flight' hormone. In the wild for animals to survive they either need to stand their ground and fight or, usually more sensibly, RUN AWAY! Nowadays humans don't often need to run away from a lion or a bear but adrenaline still kicks in in plenty of other threatening situations: at exam time, when you have to walk past the school bully, when you're taking the all-important penalty shot that could win the game, and of course, when your mum shouts for you and includes your *middle* name. You know you're in trouble then.

Imagine how you feel when you're frightened. Think about some of those scary dreams we all have where you wake up in the night and you're absolutely petrified. Something scary in your dream was chasing you and you couldn't quite get away. Your heart is pounding, your mouth is dry, you're all sweaty and it takes a few minutes to realise it was just a dream, you're safe at home and the only thing threatening you is the awful odour of foot fungus and wind wafting over from your smelly brother!

RABBIT FACT:

24 rabbits were taken to Australia by an explorer from Britain in 1859. In less than 100 years there were over 600 million, that's 600,000,000! This means about 16,428 new rabbits for every day!

Now imagine being a prey animal. Those dreams about being chased aren't just dreams — they're things you have to look out for every single minute of every single day. Prey animals have various ways of avoiding being eaten, depending on what sort of animal they are and also depending on what sort of animal is trying to eat them. Some use camouflage, some play dead, and some are quick runners. Some have fearsome defences like horns and some, like the skunk, just smell really bad! But lots of prey animals (such as rabbits) use the form of defence that human armies use — strength (and safety) in numbers.

Having more eyes on the lookout means someone can relax and eat.

Rabbits have a good look round before leaving the safety of the warren.

Living in a big group has lots of advantages. There are more eyes on the lookout and ears listening for danger. With more lookouts, everyone gets more time to eat, play and groom.

Moving as a big group can confuse predators because they suddenly forget which one of you they were after in all the chaos. And, quite selfishly, any one animal in the group is less likely to be eaten the more animals there are around it. Think back to your 'adrenaline rush' at the sight of the school bully. Would you rather go past him (or her!) alone or with a group of your friends? Sometimes we all need a little moral support.

Rabbits are one of the animals which love safety in numbers and hate to be on their own. They live in big social groups which can have dozens of rabbits in them. Within the big group they form smaller groups, usually of eight to twelve rabbits. It's a bit like being in a big school where you don't know everybody well, and you prefer to hang out with the people you like the most.

Chapter 2

RABBITS IN THE WILD

© RWAF

Rabbits live in big groups and can be difficult to spot because of their brown colour!

Groups of rabbits live in warrens, which are systems of underground tunnels or burrows all linked together. There are lots of ways in and out to help the rabbits escape from predators. Inside the warrens there are bigger areas for things like sleeping, hiding and grooming. Female rabbits make special nests in the warrens for giving birth and feeding their young. Rabbits dig these amazing underground hideaways themselves and can move massive amounts of earth when they put their minds to it. Their front legs are short but powerful with sharp, straight claws. They use their front legs to scrabble away at the ground when they dig. Their huge and very powerful back legs can then be used like a cross between a shovel and a broom to sweep all the loose dirt out of the way.

Rabbits' big back legs also let them do lots of other useful stuff. They can stand up on their back legs and stretch up really tall to peek over the long meadow grass or have a really good look around for predators.

Standing up tall gives you a great view!

If they feel worried or think they've spotted something dangerous, rabbits thump their back legs on the ground to warn the others to be alert. As they dart back to their burrows the flashing white of their tails also alerts the others.

One of the most famous rabbits of all time was called Thumper — he was named after his habit of thumping his back leg on the ground. He's one of the stars of Walt Disney's much-loved film, *Bambi*.

Those big back legs are also very good for escaping from predators. Rabbits can hop up to three metres in one jump and about a metre high, and move very quickly to escape predators. Sometimes though, just like us, they hop for the sheer fun of it.

Hopping is fun as well as useful!

Rabbits sometimes do a super standing leap when they're really happy. It's called a binky and it's impossible to watch without smiling. The rabbit suddenly leaps in the air and might twist its head and body in different directions. Binkies are very fast and if you blink you could miss them but they're a definite sign of a happy bunny when you do catch one.

We've said that there are lots of ways that animals avoid being eaten but there's one way we didn't mention, and that is good timing. Different predators hunt at different times of day and night but lots of them like really good light to hunt in because it helps them accurately spot where their prey is. For this reason many types of prey animals avoid coming out in the middle of the day and rabbits are the same.

Rabbits spend most of the day asleep or resting, like teenagers. They are what is known as a 'crepuscular' animal. This means they are most active at dusk and dawn. What you'll notice about lots of animals that are prey is the position of their eyes. Prey animals tend to have eyes that are placed on the sides of their heads, not at the front like in humans, dogs and cats. This makes a very important difference. Rabbits can see almost 360 degrees. This means they can see everything which is behind them and only have a very small blind spot in front of their nose. They literally have got eyes in the back of their head. Just like mums and teachers.

Chapter 2

RABBITS IN THE WILD

So, rabbits sleep for most of the day and then creep out of their burrows at dusk. They have a good look around on those big back legs with their enormous, sideways eyes ... and then what? They EAT! Anyone who's seen wild rabbits in their garden or on the side of the road on the way home from school in the deep dark winter will be pretty certain about what makes up the bulk of a rabbit's diet. That's right. It's just plain old grass. Rabbits do like to browse around for other things and do eat various other plants like dandelions and other leafy goodies but the vast majority of their diet is grass.

RABBIT FACT:

Rabbits are not rodents like rats and mice; they are called 'lagomorphs'. You can tell the difference because rabbits have two pairs of incisor teeth on the top jaw but rodents only have one pair.

RABBIT FACT:

Resting   Sleeping   Dead!

Rabbits can sleep with their eyes open so that sudden movements wake them up. This can make it quite difficult to tell if they are awake, asleep or even dead!

Young rabbits creeping out to explore and graze.

Grass and similar plants are hard to grind up and even harder to digest so animals which live on these foods have had to evolve ways to get at the goodness locked inside those woody cells. Firstly they need good teeth.

Rabbits have very sharp incisor teeth, which are like little chisels. There are plenty of vets who have ended up on the wrong end of these teeth and they are very painful! The incisor teeth snip through plants like garden shears, and then the food is pushed to the molar teeth at the back to be ground up. One very important thing to know about rabbit teeth is that they never stop growing throughout the whole of the rabbit's life. This is because grass and plants need to be smashed and ground really well before being swallowed. These foods don't contain many nutrients so animals living on them need to eat A LOT to get everything they need. This means rabbits have to spend most of the time they are above ground grazing, and all that grinding wears teeth down pretty quickly. This is why their teeth must keep on growing; if they didn't, they would soon disappear altogether!

Now I'm going to tell you something about rabbits that is going to make you very glad you're a human and not a rabbit. There is something else they eat besides grass and other plants ... POO! That's right, the way rabbits get all the nutrients out of their tough diet is by eating it *twice*. Eating poo is called coprophagy, and for rabbits it's simply essential. When rabbits are in their burrows they produce soft, sticky poo called caecotrophs which is covered in mucus (a bit like snot). The rabbit eats this snotty poo straight out of its own bottom. Yum! The mucus protects the poo from all the strong acids and juices in the stomach. Then, when it gets further down the intestines, it is digested a second time and all the vitamins, minerals and goodness get a second chance to be absorbed.

"eat up then, darling"

So, now that you know pretty much everything there is to know about rabbits in the wild, we had better get down to the tricky business of keeping them as happy, healthy pets.

Chapter 3

We've already said that we should try to give every one of our pets a great life. The best start to that is to get the basic survival stuff exactly right from the very beginning. The top three things needed for life are air, water and food. Animals, including humans, can't live without these things and when it comes to food, getting the right diet and feeding the right *amounts* of food will get your rabbits off to a brilliant start for a healthy and happy life.

Let's tackle the easy part first – water. Water is absolutely essential for every living thing on Earth. For animals like humans and rabbits, after the need for air, water is the most important thing. If animals can't get to enough water they can get very ill and die really quickly. Water is the only thing your rabbits need to have to drink. Depending on the food you give them they will also get some water in their food – say, for instance, in some juicy grass – but it's essential they have access to plenty of fresh water all the time as well. Just like us they will need to drink more during warm weather compared to cold weather and also if they've been very active, running, hopping and playing.

Lots of you will have seen the water bottles that most small animals have attached to the sides of their cages. These are fine but it's very important that you check them at least twice a day because the little balls that let the water flow out of the nozzles can get stuck. If this happens it might look like your rabbits have plenty of clean water, but they actually can't get at it! This would be very frustrating for them but is also very dangerous because water is vital for life. It's best to offer one or two bowls of water as well. If you think back to our wild rabbits, they are never going to drink water out of a bottle, are they? They will sip from puddles, streams and lakes depending on where they live. Lots of rabbits find it much easier and more enjoyable to drink from a bowl rather than a bottle. Talk to the people at the shop about which bowls are best and not likely to tip over and make everything cold and wet.

However you decide to give water, whether using bottles or bowls, you need to check the water at least twice a day. Make sure it is clean and there is plenty available. Bowls can end up with bedding and food in them so need to be cleaned every day, and, if they're dirty again, twice a day. Bottles might keep water clean for longer but it's still important to check them twice a day to make sure the water is flowing. You still need to change the water in the bottles every day even if it looks clean because it can get stale or get algae growing in it which can make your rabbits poorly.

So, first we need air, then water – and then of course we need food. Bodies are amazing machines and make any computer or machine humans have made look like the most ridiculously rubbish toy imaginable. Like all machines bodies need energy to make them work. Because our bodies are so spectacularly clever they can also fix themselves and replace their own parts. Instead of plastic and metal bits or replacement batteries they need vitamins and minerals, fibre, proteins, fats and carbohydrates. All animals get these from their food and over the millions of years that animals have been on the planet they have evolved different ways to get all the things they need from their foods. As we said about wild rabbits, they eat lots of plant material that doesn't have much goodness in any one bit. This means they have to eat lots of it, they need to chew it really well, and they eat it twice to make sure they get every scrap of goodness out of it.

As I said, nature is very clever at balancing things and keeping everything happy and it's usually humans that make a mess of it. This is what happened when we started keeping rabbits as pets. We thought it would be much easier and better for them if we made them some nice easy food to eat instead of all that rubbish grass. They could get all their nutrients out of a convenient bowl in easy-to-eat cereal pieces, a bit like the muesli that some people like for breakfast. There'd be no need for much of that boring grass or hay. Do you remember the danger of assuming animals are like humans? Exactly. Animals like rabbits love eating grass and hay and plants. That's what they spend most of their time doing when they are awake. Their teeth never stop growing so they absolutely have to keep chewing and chewing to keep their teeth healthy. Overgrown teeth can be disastrous for rabbits, causing pain and abscesses full of pus in their mouths and even killing them. Our idea of a good diet for them wasn't so good after all.

© PDSA

Rabbit muesli should NOT be fed to your rabbits.

Giving rabbits muesli-style food is like you being given a bowl of cabbage, bread, beans, brussels sprouts and sweets mixed together for all your meals. There are things in there that would definitely get left! Rabbits, just like you, will pick out their favourite bits — which means they won't get a balanced diet and might not get some of the vitamins and minerals they need.

The food crumbles easily as well, so they don't spend enough time chewing to keep their teeth healthy and can end up having pain, lots of trips to the vet, and even sometimes being put to sleep. On top of this, lots of them get way too fat. This happens partly because sometimes people give them more food than they need but also because they're used to eating food that's a bit rubbish! Their bodies have evolved to be experts at getting every scrap of goodness out of their food. So when humans come along and give them food full of energy in small parcels they can easily get way too fat.

FRESH WATER AND THE RIGHT FOOD.

Being too fat (or 'obesity', as it is called) is a real problem for animals and humans alike. You don't find fat animals in nature. Some animals will build up stores of fat to keep them warm and give them energy through the winter but you will never find a truly fat animal in the wild. Being too fat can give lots of animals diseases and the extra weight puts strains on joints and bones and gives the heart too much work to do. You can imagine that a fat prey animal might struggle to get away from a predator and wouldn't live long enough to have fat babies. This really is survival of the fittest!

Being overweight can also make it very hard to keep clean. Lots of animals need to groom themselves to stay clean and keep their fur and skin healthy. Being too fat makes this really hard, and for rabbits especially this is a big problem. We'll look at these problems more in Chapter Seven, but for now all we need to know is that the right diet and the right amount of food is really important.

So, muesli is out of the window for rabbits. (It can be really good for humans, though, so if your mum insists you eat it for breakfast I can't help you there!). These days we still give rabbits a little bit of dry 'complete' food but it's best to buy pellets or extruded food, NOT muesli. Pellets are made in a way that doesn't let the rabbits pick and choose which bits they like and makes sure they get everything they need. Extruded foods are the same but usually taste a bit nicer. Your rabbits only need about an eggcupful of these foods each a day and it should be by far the smallest part of their diet.

"Well, this is going to be easy!"

© PDSA

Pellet food stops selective feeding.

© PDSA

Extruded food.

First of all, keep things simple and think back to the wild. Just like their wild relatives, your rabbits need lots of what's called long fibre foods to keep their teeth and guts healthy and that means fresh grass and good quality hay. A good rule of thumb is that your rabbits need about their own body size (not weight!) of hay every day, but an unlimited supply is best.

© RWAF

You can never have too much hay!

Always make sure hay is given in a hay rack or something similar to keep it clean.

© RWAF

Hanging baskets make great hay racks.

Just having hay as bedding isn't good enough because it could be too wet and dirty for the rabbits to eat. If they have safe access to grass all the time then don't worry too much, they won't go hungry – just make sure hay is always there if they want it. Sometimes they might not feel like going outside if they are frightened or if the weather's horrible, and if they always have hay they can choose what they want to do.

© RWAF

If you curl up in your food you can eat with zero effort!

© RWAF

Rabbits in a shed enjoy some freshly picked grass.

Fresh grass is lovely for rabbits but don't give them clippings from the lawnmower. These can go off very quickly and make your rabbits ill.

FRESH WATER AND THE RIGHT FOOD.

Different leafy plants give a healthy variety.

In the wild, rabbits nibble lots of different leafy plants so a good variety for your pet rabbits is also important. Lots of plants (like dandelions, clover and nasturtiums) are safe to give. If you've rabbit-proofed your garden you need to make sure you know which plants you have growing there that the rabbits might get to. This is important because some plants (like foxgloves and ragwort) can be very poisonous. There are lots of plants of both types so finding out which is which will be part of your fact-finding research later on.

An occasional carrot is a tasty treat.

Lots of rabbits love fruits, because they are sweet, but this can be a problem. Fruits are the rabbit equivalent of a pick 'n' mix; high in sugar and irresistible! They can make your rabbits fat and cause tooth problems so should be a very occasional treat at the most.

Try and make sure that grass and hay is about 80% of what your rabbits eat. 15% should be other leafy plants and vegetables. Make sure you vary the plants they are given to keep the diet balanced and interesting. Try to avoid too many vegetables like carrots because they are full of energy and can make your rabbit fat.

The egg cup amount of dry food each will make up the last 5%. Wow. Maths as well as rabbit facts. Won't your teachers be pleased!

You might be surprised to find out that a bowl is not the only way you can give your rabbits their food.

Sometimes just having a bowl causes a bit of a jam!
© RWAF

There are lots of ways to make eating more interesting and keep your rabbits busy. This is part of what we call 'environmental enrichment'. It's a bit like you having posters on your wall, books to read or a TV rather than a blank, white box for a bedroom. So now we had better look at the type of 'environment' that keeps a rabbit happy and healthy as well as how we can 'enrich' it or make it even better.

Chapter 4

Repeat after me a thousand times:

'A HUTCH IS NOT ENOUGH!'

Well, that was a short chapter, wasn't it?

On second thoughts maybe we *should* say a little more than that.

Rabbits love room to run and play.

A couple of centuries ago when humans started keeping rabbits they were not being kept because they were cute pets. They were being kept to be eaten. In those days most people didn't really care very much about animal happiness, they just needed the animals to survive long enough to get fat enough to go in a stew! The easiest way to keep rabbits was in cages and the smaller the better so they didn't take up too much room. Unfortunately, when rabbits started becoming pets instead of dinner no one really thought to change the cages, or hutches as we now call them.

This means that for years and years people have assumed that rabbits don't need much room at all; that they can can sit in a hutch all day being fed muesli and will be perfectly happy. Actually, what it's meant is that thousands, if not millions of rabbits have had the most miserable, boring, lonely and painful lives imaginable. So how do we work out what might make our pet rabbits happier? That's right! Look to the wild.

As we said, wild rabbits live in big systems of tunnels or burrows called a warren. They are dark, safe and secure and go a very long way. When the rabbits choose to go outside they can hop, run, dig, stand up on their back legs to look round, graze, groom, binky and PLAY! Wild rabbits will often cover an area more than the size of a football pitch every day, and love to explore. So if you were thinking that rabbits don't need much room, it's time to think again. The 'normal' hutches that you see in most pet shops don't let rabbits move much at all and are way too small.

The first thing to remember is that the hutch should never be the only place your rabbits have to live. It should be somewhere that they can shelter in bad weather or hide away from things they are afraid of and they definitely need to be able to get in and out of it whenever they want to. When they choose to be in there they need to be able to stand up to their full height and stretch out to their full length to relax. They should have enough room to do three hops in a row. Small hutches that don't allow all these movements can not only make your rabbits really sad, they can cause back pain and deformities and make them very bad-tempered.

Rabbits love to stretch up tall to look around.

A shed makes a huge, luxury hutch.

Your rabbits will also need room for their food, water, litter tray or toilet area and their sleeping area. On the odd occasion when they've had enough of each other, they will need to be able to have a bit of space to themselves. All this added together means that the *smallest* hutch you should ever consider buying needs to be at least two metres long, at least 70 cm wide and at least 70 cm high. If you're thinking about large or giant breed rabbits it will need to be even bigger. Quite often in life you'll hear people say that size doesn't matter, but when it comes to rabbit hutches it definitely does matter! The bigger you can afford the better. In fact, lots of people even use garden sheds, which can make brilliant-sized rabbit houses.

THE NEED FOR THE RIGHT ENVIRONMENT.

If you do decide to use a shed, remember that they can get really hot – your family will need to make sure there is plenty of ventilation, especially in the summer. Of course, with any housing the opposite is also true and your hutch will need to be protected from the worst of the weather as well as the sunshine. Rain and wind often have a favourite direction so try and work out what that is where you live and make sure your hutch is sheltered from it. In the height of winter you'll need extra covers and draught protection and plenty of good, warm bedding. You can also move your rabbits into the garage in the winter if you have one but you'll need to leave the car outside in case the fumes make your rabbits ill – and the rabbits will still need to be able to come and go into their exercise area.

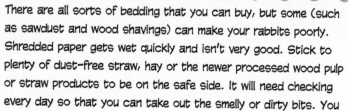

There are all sorts of bedding that you can buy, but some (such as sawdust and wood shavings) can make your rabbits poorly. Shredded paper gets wet quickly and isn't very good. Stick to plenty of dust-free straw, hay or the newer processed wood pulp or straw products to be on the safe side. It will need checking every day so that you can take out the smelly or dirty bits. You need to top up the cosy bits, especially in winter, and if you're using edible bedding like hay. At least once a week, all of the bedding should be changed.

So now you have a huge hutch or a shed which is practically a rabbit palace. Next, you'll need an exercise run. A mistake that lots of people make is to have a run that is separate from the hutch. They take the rabbits out of the hutch, carry them to the run and put them in there for however long they are at work or think the rabbits might like to be there. But how do we know what the rabbits want to do? We can't read minds. If you were in the living room and your mum was insisting on 'just quickly watching the news headlines' you might decide to go into the garden or sit in your bedroom and read a fantastic pet care book. Whatever you did would be your choice — and our pets need to have choice too. Rabbits need to be able to come and go from the hutch whenever they want to.

Imagine if you left your rabbits in a separate run and went out for hours. It could tip down with rain, or a big hungry cat might be lurking about scaring the rabbits out of their wits. With freedom of choice, they can feel safe and secure and sheltered whenever they need to or explore and play when the fancy takes them.

Remember too that rabbits are most active at dusk and dawn, so if we put them outside all day and in their hutch all night it is not natural for them at all. It would be like your teacher coming to your house, waking you up at midnight and making you do history homework for two hours. Yuk!

The other problem with carrying rabbits back and forth to a run (which might surprise you) is that rabbits don't actually like being picked up and carried about. Remember, they are prey animals and they like to know they are close to the ground and able to bolt to safety at the drop of a hat. Being picked up feels like being carried away by a bird of prey. They don't understand you are there to look after them and you are their friend. When you first get them they might even think you are a predator, and with some humans they'd be right! So picking them up and carrying them around should be avoided as much as possible because it can make them really frightened. If they're frightened they are more likely to scratch and bite you, not because they are being mean but because they are trying to get away to safety. Rabbits' very powerful back legs can not only hurt you, but sometimes (especially if you're not holding the rabbit the right way) if they kick and struggle they can damage or even break their backs. All in all, it's best to leave them on the floor as much as you can. If you were after a pet that was guaranteed to like a cuddle then rabbits might not be the ones for you. It takes a lot of time, patience and effort to gain the trust and affection of pet rabbits — and there are still no guarantees.

Chapter 4

THE NEED FOR THE RIGHT ENVIRONMENT.

With this in mind, the best thing to do is have a run that is always attached to the rabbits' hutch. This could mean the hutch sits inside a big run, or you could have a tunnel or doorway connecting the two. The main thing is that the rabbits can always choose where they want to be. There are lots of accessories you can buy to link different areas safely together, so you can always keep adding bits on to keep them entertained and interested and give them new places to explore.

If you think about the football pitch we said wild rabbits could easily cover, you can imagine that those small hutches aren't going to have any rabbits jumping with joy (not just because they were disappointed but because they wouldn't have room to!). The smallest exercise area should be at least three metres long and at least two metres wide — but just as we said with the hutch, the bigger the better.

© RWAF

The hutch should be permanently attached to the run ...

... but you can be inventive and use tunnels to attach the two together.

© RWAF

© RWAF

You can see the old style hutch (on the right) is TINY compared to the minimum we need to have today (on the left).

The run needs to have shade and protection on at least part of it so that your rabbits can be outside and graze even if the sun is strong or it's raining. A thick tarpaulin tied over a portion of the run works well for both these situations and should be fairly easy to get hold of.

A tarpaulin acts as shade from the sun and shelter from the rain.

Some lucky rabbits get the run of the whole garden. If you're planning to do this you need to be absolutely sure they can't escape. Rabbits can squeeze through very small holes indeed, they can dig their way out of lots of places, and they can chew to make holes bigger. You will need to look at your garden from a rabbit's eye view, not a human's! As we said before, you'll also need to be sure you know which plants are safe in your garden and which ones are poisonous so you can get rid of the nasty ones or make sure the rabbits definitely can't get to them. Remember that it's not just about your rabbits getting out; it's about other animals getting in too. Even the most secure gardens cannot be safe from some cats and none are safe from birds of prey. If you're planning to give your rabbits the run of the garden it always needs to be when you or your family are there to keep watch and keep them safe.

Now, you might have noticed that I mentioned a litter tray when we were talking about hutches and you might have thought I'd gone mad and forgotten I was writing about rabbits and not cats. Don't panic, my brain is ok (I think!). Rabbits are actually very easy to train to use a litter tray. They are quite clean animals by their nature and tend to always use the same area to go to the toilet. If you watch out for where that area is you can put a litter tray there and they will quickly get used to it. This makes it easier to keep the hutch or the run clean.

THE NEED FOR THE RIGHT ENVIRONMENT.

It also means that some people find it easy to keep their pet rabbits indoors instead of outdoors. If you're thinking about turning your palace into your rabbits' palace, there's one thing you absolutely have to do. CHECK WITH YOUR MUM AND DAD! I know that being the amazing, responsible pet detectives you are, you are going to check all of this with your family – but rabbits can destroy houses, so it's really important.

Remember what we said about those sharp, chisel teeth and powerful legs? Rabbits can chew carpets, cupboards, sofas and tables as well as anything else they fancy. They love to dig and can damage lots of surfaces when scrabbling around. They are great at jumping and love comfort so will probably check out all your furniture for the softest, warmest places to sleep — if that happens to be your mum's new (and now recently-shredded) jumper, she might not be too pleased!

© Marit Emilie Buseth

Rabbits can live indoors.

© Marit Emilie Buseth

... but they can chew and damage things too!

Most importantly, you need to remember how dangerous your house could be for your rabbits. Every electrical wire will need to be out of reach or covered so it can't be chewed because an electric shock will easily kill a rabbit. You'll also need to make sure all your house plants are well and truly hidden. There are so many different types of plants kept in houses these days that it's almost impossible to know which are safe to eat. It's best to be safe and keep them all out of the way. One more thing to consider is sunshine. Rabbits make vitamin D from sunshine just like humans and some indoor rabbits get low levels of this important vitamin. If you're thinking about keeping house rabbits, talk to a good rabbit vet to find out what, if anything, they recommend.

Right, back at the end of the last chapter we mentioned something called 'environmental enrichment'. Now we've learnt about the right sort of environment for pet rabbits, it's time to look at how to enrich it. Just as it sounds, this means to make it even better. We said right at the start that happiness is just as important as being healthy and that is what I'd love you to remember ALWAYS when it comes to pets. A big hutch is nice, like you having a comfy house to live in, but if you couldn't do fun things and see your friends as well you wouldn't be happy. Well rabbits are just the same and the next two needs are virtually impossible to look at separately. You'll see why as we go along.

In the meantime if there's one thing you remember from this chapter make sure it's this:

'A HUTCH IS NOT ENOUGH!'

Chapter 5

All the needs of animals are important, but for many animals this is probably the most important one for being happy — so we need to get it right. You might have noticed that so far in this book I have only used the word 'rabbit' a few times. 'That's bonkers!' I can hear you say, 'the book is chock-a-block with the word'. Aha! If you look back now you'll see that I always used the word 'rabbits', plural. This is because I was sneakily trying to get into your brain and make you always think about lots of rabbits, not just one. Sadly, for thousands and thousands of rabbits, humans have got it wrong for a very long time. In the past (and, I'm afraid to say, still today) some rabbits have been (and are still) kept on their own. You need to immediately wipe this from your mind because the most important thing you need to remember from this chapter is this:

Rabbits should NEVER be kept on their own.

Now there might be occasional times they have to be alone for a short while (for example, if one of them is poorly) but in general for everyday life they should have at least one rabbit friend, if not more.

Let's do what we should always do and think back to our wild rabbits. They live in really big groups, usually in smaller gangs of very good friends and family. They are prey animals and find safety in numbers.

Rabbits feel safest with other rabbits.

More lookouts mean someone can always relax and eat.

This means they feel safest when there are other rabbits around. Think back to how they communicate and work together. If there is more than one of them they have extra eyes and ears to look out for predators. If they know someone else is on watch they can relax a little and graze and feel secure. This means they can get plenty to eat and keep their teeth and intestines healthy.

Not only are they prey animals, needing numbers to feel safe, they actually really like other rabbits. They are sociable animals. They like to play together, which for pet rabbits means they will stay fitter and slimmer. They can run and jump and chase each other around — especially in that massive run you're going to have! Life should be fun for them.

Rabbits love to snuggle, just like humans do. It not only makes them feel safe and happy, but in cold weather it keeps them warm. Rabbits are small animals and small animals lose heat more quickly than big animals. This is because they have a big body surface compared to their size. By snuggling together they can share warmth and slow down the loss of heat from their little bodies.

© Marit Emilie Buseth

Rabbits love to snuggle up with their friends.

© Marit Emilie Buseth

Having a friend is useful for washing those hard-to-reach places!

© Marit Emilie Buseth

Rabbits also love to groom each other. They can do some of it for themselves but like all of us they have places that are difficult to reach or see, like their faces, eyes and ears. You can imagine that if you only have your tongue and your paws to wash with, having a friend to help is always going to make life easier. Grooming each other helps keep their fur in tip-top condition, but is also an important part of rabbit life. It helps them form relationships, bond with each other and keep the group tightly knitted together. If you do get some pet rabbits, if you have time and patience they will sometimes learn to love you grooming their heads and ears, just like their friends do.

Chapter 5

You might find it hard to believe, but we humans are quite similar to rabbits in lots of ways. Deep down inside we are still prey animals. We have lots of ways and lots of knowledge these days to help us avoid being eaten but there are still plenty of animals that could make a meal of us if we got too close. We still feel scared of things like the dark or spiders and even when we try to tell ourselves we don't need to be afraid we sometimes still are. This is because our bodies are still ready to avoid danger so we still get that racing heart and sweaty feeling when we feel under threat. You might remember when we talked about how you would feel if you had to walk past the school bully. Would you feel confident to do it on your own or would you feel safer if you were in a crowd of friends?

Like rabbits, we too are social creatures. Ok, some of us don't seem very friendly, but on the whole humans like to be near other humans. You can have fun on your own but the best laughs and happiest times tend to be with other people. We like to share experiences, swap stories and news and (maybe most importantly) play together. When we are sad, frightened or poorly we usually need other people even more. When everything seems bad, a warm hug from someone you love is usually the best medicine.

So although we said we shouldn't assume animals' needs are exactly the same as ours, we can still empathise with them. This means we can imagine how they might feel in some situations because we know how we would feel. I'm sure you can imagine how you would feel if you spent your whole life alone. You could definitely survive but you would probably at times be frightened, grumpy, sad, bored or frustrated and feel like there was something missing. To be honest, you would probably go a little bit mad.

So imagine yourself living alone in your empty bedroom. To be kind, someone decides to put a donkey in your bedroom to keep you company. (Bear with me, I am going somewhere with this!) You can't understand what a donkey 'says' and a donkey can't understand your ramblings either. If you get too close or try to cuddle him he might bite you or kick you and really hurt you. You might spend all your time squeezed up into a corner just to try and stay out of his way. This is what it's like for rabbits and guinea pigs who are kept together.

Over the years lots of people with all the best intentions have kept rabbits and guinea pigs together – but they are totally different animals! They don't communicate in the same way, they have different needs when it comes to food, and they don't live together in the wild. They can both hurt each other with bites, but rabbits are like the donkey in your bedroom. They are much bigger and much stronger, and can badly injure and even kill guinea pigs. Please don't keep them together. If you like guinea pigs, find out all about their needs and learn how to keep them happy and healthy with other guinea pigs, not with rabbits. In the meantime, let's concentrate on the bunnies!

Exploring is more fun with a friend!

Now that you're really empathising with those lonely rabbits which are kept on their own, imagine how happy they would feel if they had a proper rabbit friend or two.

If you or one of your friends already has a lonely rabbit, don't worry — all is not lost. You can nearly always introduce another rabbit and if you do it in the right way they will usually become firm friends very quickly. Lots of animal rescue centres have rabbits looking for homes as well as all the cats and dogs. By adopting a rabbit, you or your friend will be doing a double good deed; giving two rabbits a happy life with a friend and donating some money to a worthy cause at the same time. Talk to your vet about the best way to introduce them.

One thing you need to remember while we're talking about keeping rabbits together is how good they are at having babies! You might think baby rabbits are very cute, and of course you'd be right — but they soon turn into adult rabbits and have even more babies.

Hey! Are you coming out to play?

Now that you know how much room you need for even two rabbits, you can imagine that letting your rabbits have babies is not really the responsible thing to do. With this in mind you'll need to have your rabbits neutered. We'll talk about this in Chapter Seven. For now, though, all you need to know is that the best pair to have is a neutered male with a neutered female, or for a group, a mix of male and female neutered rabbits. Pairs of the same sex can find it harder to get along so are best avoided.

You might be thinking that I keep saying I'm going to talk about environmental enrichment and then forgetting about it. I haven't forgotten at all. Keeping rabbits in pairs or more is the best possible way to enrich their environment! They can play, chase, watch out for each other, keep warm and groom. What we're actually saying is that they can suddenly do loads of natural behaviours which they not only enjoy but that they feel the *need* to do. And this leads us on to the rest of enrichment and what it's actually all about. It's about another rabbit need: the need to behave naturally. And I'm sure now that you can understand why an animal's social needs are so entwined with behaviour. If you're a social creature you want to do fun things with your friends!

Chapter 6

Have you ever had an itch in a place you couldn't reach? Right in between your shoulder blades, for example? You bend your arm right the way up your back but the tip of your thumb stops about a centimetre short of the itchy spot. Ooooohh ... it drives you mad, doesn't it? You try and ignore it but the more you try the more you keep thinking about it until you can't think about anything else. Eventually your mum finds you crazily scraping yourself up and down the door frame, or discovers that you have roped her best hairbrush to a wooden spoon and are frantically gouging it up and down your back, with a weird look on your face like a cross between sheer panic and total heaven.

This is what it's like for animals who are not allowed to do the things they love or feel the need to do. It's the itch they can never scratch. All animals are born with some behaviours that are what's called 'innate'. This means they are born needing to do something even if they don't know why. Other behaviours are learnt as they grow. For example, an innate behaviour for children appears to be constantly picking their noses, whereas a learnt one is getting a tissue and actually wiping it!

Innate behaviours help animals get a head start because they can do things without needing to be shown. One of the strongest and earliest innate behaviours you see is when animals suckle their mother's milk. Within minutes of being born, calves, lambs, kittens, puppies and human babies all start looking for their first warm drink of milk. They don't think about why (and they don't need to), but it gives them a great start because they get a full tummy and lots of goodness straight away.

The important thing to remember is that even if we keep an animal in a way that means it doesn't *need* to do something any more, it will still feel the urge to do it. It will still be the itch they can't scratch. A perfect example for rabbits is digging.

We know very well by now that rabbits in the wild live in burrows which they usually dig themselves. When we keep them as pets we get them a nice big hutch or a shed to live in, so you might think they will realise we've saved them the trouble of all that digging. But you'd be wrong. Rabbits love to dig, they feel a need to do it that they can't explain — and not just because they can't talk!

Digging is something that the majority of pet rabbits are never allowed to do. People don't often appreciate a massive hole in their lawn. It can also be dangerous for the rabbits if they tunnel out of their run or garden because they might get lost or get caught by a predator. So lots of owners put wire in all sorts of places to stop them digging. This is all very well – we want them to be safe – but you still need to find a way to let them have the enjoyment of digging. We'll come back to that later.

Our pets should be able to have as much fun and be as happy as possible. Think back to all the things we've mentioned time and again in the book that rabbits enjoy. Digging, grooming, jumping, exploring, hiding, hopping, standing up, leaping in the air, grazing, relaxing, cuddling — the list goes on. So now let's look at all the ways we can make it happen for them. Or in other words, all the ways we can enrich their environment. See, I haven't forgotten!

© Marit Emilie Buseth

A friend is the best environmental enrichment there is.

We know after the last chapter that they need to have a friend or two, and that's definitely going to tick some things off our list. They can groom and cuddle and play.

© Marit Emilie Buseth

A hop, skip and a jump for joy.

You've hopefully learnt that a small hutch is not going to do much for them, and that with a big hutch and an attached large run they can run and hop, stretch up, and jump.

THE NEED TO EXPRESS NORMAL BEHAVIOUR.

© RWAF

Hanging baskets make great hay racks.

But even the biggest run is still going to be pretty small compared to that football pitch we mentioned earlier, and it will be pretty samey every day — so you need to make it interesting and you need to change it around once in a while to keep it interesting.

Hopefully your run will be on some grass so they can graze and maybe even forage around for some other plants as well as grass. You'll be giving them some vegetables and other leafy plants every day too. It's also good to have hay available for them all the time. This might be in their hutch but you can get little hay racks for the run too so they have plenty of choice. The run and their environment is your chance to be really creative. Your mum or dad might have an old hanging basket. You could dangle it from the roof of the run to use as a hay rack and keep the hay nice and dry.

Don't just give them their veg or pellets in a bowl. Have a look in the pet shop or be inventive at home. Hide their food in cardboard tunnels or scatter it among some grass or hay so they can explore and enjoy hunting around to get their food.

We've said they live in burrows and that these are where they feel safe, so try making some tunnels for them. Cardboard tubes or boxes with holes are an easy way to give them places to hide and explore. You can buy tunnel systems from pet shops if you have enough money. You can even make tunnels leading from the hutch to the run or from the big run to another smaller run somewhere else. Challenge yourself to see how inventive you can be.

As well as hiding places, rabbits also like look-out posts or platforms. You can put things like a bale of straw in the run so they can hop on and off. Lots of things can work. You can use a strong box or a wooden crate. If your rabbits had a small hutch when you got them, you could stand that in the run as a place for them to explore or sit. Hopping up and down is good exercise and will help to keep them slim, but will also keep their bones strong. When they are on top of the platform they can have a good look round or just relax — it's their choice!

The great thing about environmental enrichment is that you can let your imagination go wild. There is nothing better in the world than seeing a happy pet and you'll soon see when you've made them happy.

Let your imagination go wild!

THE NEED TO EXPRESS NORMAL BEHAVIOUR.

Pet shops sell lots of toys for rabbits, so go for it. Don't overwhelm them, though; it's important to keep things interesting. You know yourself how soon a new toy can lose its appeal and how exciting it is if you're lucky enough to get a new present, so keep that in mind. Put a couple of toys in the run but change them every few days. You can always put the old ones back in again the next week. They'll still be interesting after a break.

You can try all sorts, as long as you're putting things in that are safe. Why not give them a sturdy ball to nudge about? You can buy balls to put their pellets in with holes to let the food drop out as they nudge them around. This is way more fun that just giving them their eggcupful in a boring dish.

Rabbits in the wild sometimes chew bark from trees as well as grazing, and you can get lots of different chews for your rabbits in all sorts of shapes and sizes. You don't need to spend a fortune. You could find out what the pet shop sells and then see if you can do a home-made version. You might have some logs at home that your mum or dad could chop up for you. Just make sure you know what wood you have so you can find out if it's safe for your rabbits, and don't use wood that's had chemicals put on it. One of the behaviours that most kids love to do is making things. Why not combine your need for fun with your rabbits' needs and see how spectacular you can be? You can always send your pictures in to www.thepetsite.co.uk to compare with the other creative pet detectives out there.

© RWAF

Balls which drop food can make pellets more fun.

So what about that tricky innate behaviour we mentioned right at the start — digging? Digging might seem a difficult one to help with, but it's our responsibility as their owners to make sure they can do it. How about an old planter filled with soil from the garden? If your mum and dad are willing, you could always sink a small pond liner, a plastic bucket or a planter in the garden in the run and fill it with soil. This way your rabbits can dig as much as they like. You'll need to make sure it's underneath a sheltered part of the run so it doesn't fill up with rain water, or make some holes in the bottom for drainage. It might be messy but that's why doing your research is the right thing to do. If your family doesn't want the garden or the house taken over by a gang of muddy rabbits then you might be better off thinking about a different pet! It doesn't have to be too messy. You can use lots of different things for them to dig in, like some play sand — maybe even in an old sand pit if you have one.

© RWAF
Your parents might not want this sort of mess in the kitchen

These behavioural needs are very important for the happiness of your rabbits, and you might need to consider them even more if you're thinking about keeping your rabbits indoors. They're still going to need look-outs, hay, places to explore, hiding places, friends and space to run and jump. And they still need to be able to dig. Even if they don't have soil to dig in there could still be quite a bit of mess and your family needs to be sure they can put up with it and also keep the rabbits safe.

Whenever you think about the happiness of your pets, I think it's good to think about your own happiness. Always remember that surviving is not enough; being happy is just as important. Always try and think about how you would feel if you couldn't do the things you really enjoy or see the people you like and love. It's not ok to deprive an animal of something it needs just because we don't like it or it's inconvenient for us. Remember we said that you should never ask yourself what sort of pet you want, only what sort of pet you can keep truly happy and healthy.

Blimey, I went all serious again then, didn't I? Well, it's good to be serious sometimes because as we said, looking after animals is a serious business! So now you fabulous fact-finders know EVERYTHING there is to know about wild rabbits and how to keep pet rabbits *happy*, we had better get down to that other very important bit where we find out how to keep them *healthy* too!

© RWAF
A large litter tray full of soil is great for digging in ...

... but even a plant pot will do.
© RWAF

Chapter 7

THE NEED TO BE PROTECTED FROM PAIN, INJURY AND DISEASE.

Every person and every animal gets poorly or hurt from time to time — that's just a simple fact of being alive. I always tell my daughters that their bumped shins and skinned knees are a good sign they've been having plenty of fun. We talked earlier in this book about what amazing machines bodies are. One of their most amazing abilities is how they can heal and recover from injuries and disease. But you'll all know already that there are lots of illnesses and injuries that bodies can't cope with and will need extra help to recover from. It's not just about getting help for your pets when they are poorly or hurt; it's very important to find out all the ways you can stop them from getting ill in the first place.

We said earlier that rabbits don't like being handled much and carried about. However, it is also important for health reasons to make sure your rabbits are used to being handled. You will need to check them over regularly to make sure they are healthy, and they will also sometimes need to see the vet. If they are used to humans and to being carefully picked up and examined they will be less stressed and less likely to lash out and injure you or themselves.

Remember that rabbits are prey animals, so if they are picked up from above they will feel like they've been caught and will be more likely to struggle to get away. Always approach quietly and carefully from the side. Slide one hand under the rabbit's chest and use the other hand to make sure its bottom is supported. Never let rabbits dangle or use their ears when you pick them up. It's a good idea to go to see your vet as soon as possible after you get your rabbits. They can have a good check over and you can watch and learn how best to handle them. If you are unsure, always get a grown-up to help you.

Let's look at prevention first, and then we can have a look at signs to watch out for that might mean your rabbits are under the weather.

Vaccinations

Depending on your age you may or may not remember having your own vaccinations done. Having them done might involve the slight niggle of a jab but they are one of the biggest life savers of recent times. Once humans started to understand how our bodies fight off diseases we realised we could help out with some of the most horrible ones.

Your body and the bodies of most animals have a super army inside them called the immune system. This army is made up of millions of cells which are always on the lookout for bad bacteria and viruses. They never sleep, even when you do, and they happily lay down their lives to kill the bugs which try to make you poorly. You can think of these cells as your soldiers. They all have different ways of dealing with invaders. Some engulf the germs and eat them and some cells use bullets called antibodies. These soldier cells lurk round your body and when they find an invader they attack it. Now, the first time they meet a new invader they won't have exactly the right sort of bullet to instantly get through the invader's armour, so usually it takes a lot of time and effort from the soldiers to try to overpower the invader and find out what their weakness is. With some nasty diseases, the invaders are too strong and our immune cells can't win the battle and don't get the chance to make the right antibodies. These are the diseases which, especially in the past, made us and our pets really ill and even sometimes die.

Vaccinations give your soldiers super, armour-piercing bullets. Humans have studied the viruses and bacteria from many killer diseases. To make a vaccination they get some of the particular bug in a laboratory and they take away its weapons so it can't make you properly ill. The vaccine of weakened bugs is injected into you. Your soldier cells find the invaders and work out how to design the right bullets or antibodies. They kill the weakened intruders and they store away the design for the right antibody bullet. This means that when your body gets attacked by the real bug with all its weapons, your immune system is ahead of the game. Your soldier cells immediately make the exact bullet to get through their armour, and it's game over before the war even begins. Millions and millions of humans and animals owe their lives to the power of vaccinations.

Different animals get different diseases so they usually need very different vaccinations. This is why you need to find out all the facts for all the animals you're thinking about owning. Rabbits need vaccinations against two killer diseases. These are called myxomatosis and Rabbit Haemorrhagic Disease (RHD).

Myxomatosis

Myxomatosis, or myxi for short, is a really horrible disease. It was actually used deliberately by humans to try to kill rabbits when they became such pests in Australia. The virus was taken to Australia in 1950 and killed millions of rabbits. It has since spread through much of the world and is also a big killer of pet rabbits. The main signs you see are very swollen eyes and bottoms. The rabbits have runny eyes and noses and can have big lumps all over their bodies. As they get more poorly they stop eating, and usually die in one to two weeks. It's a very nasty way to go.

A rabbit with a milder case of myxi, with skin lumps developing.

A rabbit who has died from myxi. His eyes are swollen tightly shut and his fur is wet from the discharge.

© RWAF

Wild rabbits' immune systems have got a little bit used to the virus over the years but pet rabbits don't have any immunity to it. This means that nearly every pet rabbit which catches the disease will die — unless they're vaccinated! It's very important to talk to your vet about when to vaccinate and how often. Virtually all vaccinations need repeating to jog the memory of your soldier cells about which antibodies to make, and myxi is no different — so make sure you keep your rabbits' boosters up to date. Your vet will be able to tell you how often you'll need to have all your pets' vaccinations repeated, and they'll usually be able to send you reminders to jog your memory too!

Now, myxomatosis is a tricky customer and it can change its weapons now and again. This means that occasionally a vaccinated rabbit can still catch the disease. The good news is though that virtually all vaccinated rabbits survive if they do get it and usually only have mild signs.

Myxi is spread by biting parasites like fleas, mosquitoes and mites. These creatures accidentally take the virus in when they bite an infected rabbit. The next time they hop onto a new rabbit and nibble that one, the virus is passed on. It can also be spread directly between rabbits. So as well as vaccination, if you can keep the numbers of insects and mites down then you can help make it less likely your rabbit will get the disease at all. Insect screens on hutches and runs help, as does making sure there's no standing water (like rain butts and ponds) in your garden — mosquitoes love to breed in these places. Be VERY careful if you are using insect repellents or treatments to kill insects and mites. Lots of the things we use for dogs and cats are very poisonous to rabbits and you could make them very ill or even kill them if you use the wrong thing. Make sure you check with your vet before you put any chemicals or medication near or on your rabbits.

Obviously indoor rabbits might be at less risk but they can still get the disease. You might end up bringing rabbit fleas back on your clothes from a walk in the park, especially if you've got a dog or cat that chases rabbits. I'm sure you've all seen the odd pesky mosquito in your bedroom at certain times of the year too, so indoor rabbits are not guaranteed to be safe at all.

ALL pet rabbits should be vaccinated against myxomatosis. And that's all there is to say about that!

Rabbit Haemorrhagic Disease (RHD), Viral Haemorrhagic Disease (VHD)

These aren't two different diseases, they are just names and abbreviations for the same disease. This is another very nasty killer disease in wild and pet rabbits. The tricky thing about RHD is that often the only sign you see is, sadly, a dead rabbit. The virus kills very quickly. When you do see signs there could be all sorts, from fits to runny eyes to paralysis (paralysis means not being able to move properly or at all).

RHD can be spread from rabbit to rabbit but it can also travel on the wind and on anything an infected rabbit has touched. This could be bowls, bedding, hutches or even your clothes if you have touched an infected rabbit without knowing.

Just the same as with myxi, by far the best way to make sure it isn't a problem for your pet rabbits is to have them vaccinated. You can get a vaccine now that is for myxi and RHD together which means fewer trips to the vet for your rabbits. Remember, though – you need to make sure you know when your rabbits' boosters are due and keep them up to date.

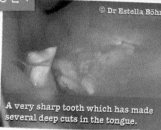

Tooth Problems, Fat Rabbits and Fly Strike

Now, these sound like very different problems, don't they? Well, they are – but they also have one huge thing in common, so I think we should look at them all together. If you think back to some of the things we said in Chapter Three, you might have an idea what it is. That's right: DIET! So, if you've taken on board all of Chapter Three, I won't need to say too much because you are already ahead of the game.

Tooth problems used to be a huge issue for pet rabbits because we didn't understand how important grinding and chewing were along with the right diet. And obviously today, if we get their diet wrong, tooth problems can still cause pain, abscesses and even death. We said before that rabbits' teeth never stop growing because they have evolved to eat grass and woody plants which need lots of chewing. If you feed things like muesli or lots of soft, sugary fruit and not enough long fibre like hay and grass, your rabbits' teeth will not get worn down enough. This means they will get too long and will start to curl in all sorts of directions in the rabbit's mouth.

They will get very sharp and grow into all the soft, tender bits of the rabbit's mouth, like the tongue, lips and cheeks. If you've ever bitten your tongue or cheek by accident you'll know how sore this is, so imagine if you had that pain all the time.

A very sharp tooth which has made several deep cuts in the tongue.

A very long incisor.

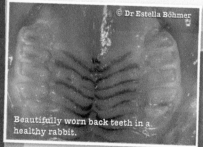

Beautifully worn back teeth in a healthy rabbit.

Spot the Difference!

Overgrown back teeth, out of line with sharp edges.

Spot the Difference!

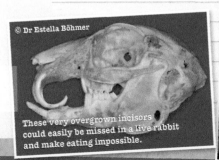

These very overgrown incisors could easily be missed in a live rabbit and make eating impossible.

... looks even longer when you look more closely.

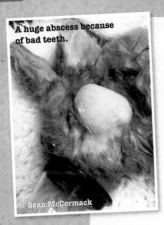
A huge abscess because of bad teeth.
© Sean McCormack

Often the bad diet that lets the teeth get too long also doesn't have all the right vitamins and minerals and this can cause weak bones and tooth sockets. The long teeth start to get loose and infection can get into the pockets the teeth sit in causing big, painful abscesses full of pus in the jaw and gums. Ouch!

The rabbits won't be able to eat and will quickly get weak and really poorly. Once the teeth get this bad it can be almost impossible to make them right again and many rabbits like this end up being put to sleep – so it's very important to stop it happening in the first place. You won't be able to see your rabbits' back teeth so if you have any worries get your vet to check them for you.

Thinking back, you'll remember we said that animals being too fat is very bad for them. Being too fat (obese) puts extra strain on joints, bones and the heart, and shortens life. It can cause quite a few diseases in various animals, and in rabbits especially it causes some very big problems. Now you probably felt a bit sick when you thought about eating your own poo all the way back in Chapter Two, but for rabbits it is essential. As we said, they have to digest their food twice to get all the goodness out of it. To do this they eat one type of soft poo straight from their bottom. Then the next time it comes out it is the hard, shiny rabbit droppings that we are used to seeing. If a rabbit is too fat it can't even

reach its bottom in the first place. This means two things. First, the rabbit misses out on some goodness because the poo doesn't get eaten. Second, the soft, sticky poo stays round the bottom and gets stuck in the fur and on the skin, and also makes the bedding and hutch a pooey mess.

© RWAF
A horribly fat rabbit compared with ...

© RWAF
... some beautifully slim ones.

Sticky poo stuck to a fat rabbit's bottom.
© RWAF

I'm sure you can imagine that being covered in its own poo is not very nice for any animal, but it can also cause something far more serious for rabbits; something called fly strike. Some flies love to lay their eggs in poo as part of their life cycle. The eggs hatch out into maggots and the maggots are very hungry. Sadly for pooey rabbits the thing the maggots are hungry for is flesh. The maggots crawl through the poo and onto the skin, and they start to literally eat the rabbits alive. I don't need to tell you how horrible this would be for the rabbits.

Maggots hatch out in the poo and start looking for food.

© RWAF

Now, you might think that fly strike would be totally obvious and that you'd notice straight away, but you'd be wrong, because there is another fact about some prey animals that we haven't mentioned yet; they don't show pain like we do. If you or I (or a predator like a dog or cat) had a big cut or a broken leg or we were being eaten alive by maggots we would be howling, crying, limping and generally carrying on as much as we could to get sympathy and help. Not rabbits. Cuts and poorly legs and injuries in general make prey animals much more likely to get caught and eaten because they are not fully fit to get away. Because of this lots of prey animals will not cry out or limp because this makes them so much more likely to attract a predator. Your pet rabbits still have this innate behaviour even though they have you to look after them. Very often one of the signs of the rabbits who are the most ill or the most in pain is just that they are very quiet and don't move around much.

Sadly, lots of people don't look at their rabbits often enough to notice these things and by the time the fly strike is discovered it can be absolutely horrific. So how can we prevent it?

DIET! The right diet and exercise will keep your rabbit healthy and slim so it shouldn't get covered in poo. You'll also need to check the hutch every day to make sure there is no soiled bedding attracting flies. The insect screens we mentioned before will help, and even things like sticky fly papers hung near or in the run will help too. You can also get fly repellents to put round your rabbits' bottoms, but remember to make sure you have the right thing and talk to your vet about when and how to put it on.

The most important thing to do when it comes to fly strike and keeping rabbits in general is to make sure you look at them frequently! This might sound simple but when it's a bit cold or you've got a friend over after school or your favourite TV programme is on it can easily get forgotten. You should look at your rabbits' whole bodies every day, and when the flies are about you should look at them twice a day. And this means carefully handling them and actually checking there is no poo and no eggs or maggots anywhere in sight.

Neutering

Neutering is when your pet has an operation to make sure he or she can't have any babies. This is important to help reduce the number of unwanted or neglected pets. We said earlier that rabbits are happiest living in mixed pairs or groups and that if you didn't have them neutered you would very soon be overrun with a mountain of new rabbits. So for fast breeders like rabbits, neutering really is essential. However, you might be surprised to hear that it also has lots of health benefits too.

Rabbits which aren't neutered can be quite grumpy and aggressive to you and your other rabbits, so you'll find that neutering keeps everyone happier and means you shouldn't have any fights.

When your rabbits are neutered the vet does an operation that takes away the male rabbit's testicles and the female rabbit's uterus and ovaries. Male rabbits who aren't neutered can sometimes get a disease called cancer in their testicles and also in a gland called the prostate. This won't happen if they have been neutered. For female rabbits it's even more important because cancer of the uterus is a much more common (and very serious) disease. Neutering will totally prevent it. Female rabbits can also get serious infections in the uterus and false pregnancies, both of which are also prevented by neutering.

Talk to your vet about when to have your rabbits neutered and what you need to do to help them get well soon after their trip to the surgery. Once it's done, your rabbits will have a much better chance of being happy and healthy for longer.

Are Your Rabbits Healthy?

The things we've mentioned so far are the big things that you need to be aware of when it comes to having pet rabbits – but there are all sorts of other diseases, conditions and injuries that can happen. If you want to learn about every single one then you should do your homework, work hard and become a vet, because I haven't got room for all of it here! The best way to keep your pets healthy in general is to be observant and know what's normal so you can spot when things are wrong. As we have said, rabbits can be very brave when it comes to not showing illness, so it's up to you to spot the little changes in the look or behaviour of your rabbits so you can get on top of problems quickly. Smaller animals like rabbits can get weak and ill very quickly, so if you are ever worried, take them to the vet. Your vet won't think you're silly if there's nothing wrong. It's always better to be safe than sorry.

Things to watch for in your rabbits themselves;

- Bright eyes. No runny eyes or discharge, no swelling round the eyelids or redness. Runny eyes can be caused by infections or scratches, but can also be a sign of tooth root problems.

- Clean nose. No snot or mucus coming from either side, and no lumps or bumps.

- Shiny, clean fur. Your rabbits will groom themselves and each other, but depending on the type of rabbit you have, you will need to groom them yourself too — especially when they are moulting. Grooming can help you make friends with your rabbits and is a good way to check them over. In general, check the fur has no matted areas or things stuck in it. If you see dandruff, especially round their shoulders, have them looked at by your vet. They could have mites (and rabbit fur mites, called Cheyletiella, like children too — so watch out!). Short-haired rabbits are much easier to keep tangle-free and have a much more natural coat. Long-haired rabbits are best avoided. Look out for wet fur under the chin. This could be a sign of drooling and dribbling and therefore tooth problems.

- Comfortable, clean ears. If your rabbits' ears seem dirty or itchy or you keep seeing them shaking their heads, they could have ear mites. Time for a trip to the vet.

- Body Condition Score (BCS). It's really good to understand body condition. It's a way of talking about how fat or thin your animals are. Most people have a scale of 1-5 where 1 is dangerously thin, 3 is normal and 5 is dangerously obese. Lots of people misjudge BCS so ask your vet how you can tell what the scores look and feel like on your rabbits. Keeping your rabbits in the right body condition is essential for good health.

- Make sure your rabbits don't have any sore patches anywhere. You need to check them all over, and especially their joints and feet.

- Make sure their nails are not overgrown. Depending on how much digging and scrabbling your rabbits do, they will occasionally need their nails cutting. Your vet or vet nurse will be happy to tell you if they need doing.

- Check for poo round the bottom and for fly eggs or maggots. Once every day, and twice every day in the spring and summer.

- Short, even teeth. As we said, you'll need your vet to check the back teeth, but if you gently lift your rabbits' lips at the front you'll be able to see their front teeth. These should be straight, short and symmetrical.

- Look out for odd behaviour. This could be anything from being 'a bit quiet' to hunched, not eating, not moving around as normal, stiff, floppy, teeth grinding, or sneezing. Basically, anything out of the ordinary. Rabbits can also get a parasite in their brain called E. cuniculi which can make them have an odd head tilt or even spin over and over. If you see any weird or unusual behaviour get your rabbits to the vet as soon as you can.

Things to Watch for in Your Rabbits' Surroundings:

- Normal faeces (poo) and urine (wee). Rabbit urine can be all sorts of colours depending on their diet but if you see blood in it you need to go to a vet. If you see the soft sticky faeces a lot of the time then it could be a sign of trouble. Diarrhoea is a definite sign that all is not well. Once again, you'll soon get used to what's normal for your rabbits. If you don't look, you don't know!

- Dropped or half-chewed food. If you find clumps of half-chewed food, it could be a sign that one or more of your rabbits has tooth problems.

Don't panic about all these things; the more you get to know your rabbits, the sooner and more easily you'll spot the odd things. The better an owner you are, the healthier your rabbits will be and hopefully the fewer trips to the vet you'll need. Remember that you can't get rabbits and just forget about them. You need to be around them at least once a day, check them over and make sure they are safe, protected and well. If in doubt, ask your vet.

Now you have all the facts about wild rabbits and you know how to keep your pet rabbits happy as well as healthy. So it must be time for even more fact-finding, some virtual reality and some good old maths. Basically it's time to actually answer that crucial question:

Are rabbits the right pets for me and my family?

Chapter 8

In this chapter, we're going to crunch some numbers and you're going to have to start investigating facts from some other places besides this book. We're also going to embark on a virtual month of being a rabbit owner. You might feel silly, but it's a brilliant way of checking if you actually do have what it takes to be a dedicated owner of rabbits. You can do it all by yourself or you can involve the whole family in the decisions, care and sums.

If you find yourself, towards the end of the virtual month, getting a bit bored with pretending to check over, feed, handle and groom a teddy once or twice a day, remember this: well cared-for pet rabbits can live for up to ten years, so you need to get used to it!

Week One – How Much??!

We'll ease you into this gently by spending the first week finding out some costs. Some of these costs, such as neutering, are definitely a one-off. Other costs, such as buying the hutch and run, might seem like one-off costs too, but bear in mind that over ten years some things might wear out and need replacing, so a bit of a reserve budget is always needed. It's good to rotate or renew things like toys, so try to get an idea of the cost of, say, three average toys – then you can allow for that amount every month.

Now is also the time to consider how much room you have. If you don't have the space for the big hutch or shed and run, or you've decided the house is not the right place for rabbits, then you might not need to go much further and can avoid the painful maths! Assuming you do have enough room, here's a place to start filling in those numbers. The boxes which are coloured in green are either one-off costs or are costs which will only rarely need to be repeated. This will help you and your family get an idea of how much your initial outlay is likely to be compared to your ongoing costs. Don't forget, though, that virtually everything will need to be paid for at the start, so add everything together for your start-up costs!

Things to find out from the pet shop/adoption centre or the internet

Item	Cost £
The bunnies! Bear in mind that adoption centre costs may include neutering, vaccinating and microchipping. Remember to multiply the cost for the number of rabbits you are planning to have.	
Hutch or shed. Remember that you need a minimum size of 2m x 0.7m x 0.7m for two average-sized rabbits. Bigger breeds and larger numbers will need more!	
Run or the materials to build one (minimum size of 3m x 2m).	
Tunnels if you're planning to have tunnels linking smaller exercise areas.	
Bedding. Try to find out roughly how long a bag will last so you can work out monthly costs for your mum and dad.	
Good quality hay. Try to find out roughly how long a bag will last so you can work out monthly costs for your mum and dad.	
Pellets. Try to find out roughly how long a bag will last so you can work out monthly costs for your mum and dad.	
Veg. You'll need to ask your parents this one. They'll need to buy extra veg to make sure your rabbits get a good variety.	
Food bowls or scatter balls.	
Water bowls and bottles.	
Hay nets or hanging baskets.	
Toys and chews. Find out how often chews will need replacing and get an idea of how much toys cost to allow for occasional replacements.	
Planter or pots for digging in.	
Litter tray.	
Brushes or gloves for grooming.	
Fly screens and papers for the summer.	
Total	

ARE RABBITS THE RIGHT PET FOR ME AND MY FAMILY?

Things to find out from your vet

Procedure	Cost £
Microchipping	
Neutering. This will be different for males and females so make sure you find out both.	
Vaccination against myxi (or combined). Find out how often boosters are due.	
Vaccination against RHD (or combined). Find out how often boosters are due.	
Pet insurance. (Per month or per year).	
Fly repellent.	
E. cuniculi treatment or prevention. Some vets recommend occasional medicine to prevent this disease and some prefer to treat it only if it appears. Find out what your vet recommends and why, and any costs.	
Nail trim. Find out how often they feel it usually needs doing.	
Normal consultation. Many people think rabbits are cheaper to have examined by a vet when they are poorly but most charge the same as for a dog or cat, so it's worth finding out what that cost is.	
Total from this and the last page added together	

These numbers may be a bit mind-bogglingly big when you look at them, but that's why I wanted you to do it. There is no such thing as a cheap pet. Obviously you will have worked out that once you're all set up, your monthly costs may be much more manageable — but don't forget about vets' fees and unexpected problems. The brilliant charity the PDSA produce a report every year called the PAW report which looks at how animals are cared for in the UK. They estimate that the average rabbit costs a massive £9000 in its lifetime — and remember that you need to have more than one!

Week Two — Handle with Care

This week you are going to spend time and energy devoted to your new (pretend!) rabbits and getting to know them. It's important to get your rabbits used to being handled right from a young age. As we said in the last chapter, this will help them feel safe and secure with you and not feel threatened. If they are happy being handled gently they will be much less likely to scratch or bite you and also less likely to hurt themselves thrashing and kicking to get away from you. It will also make them feel less stressed when they need to go to the vet and be handled by strangers.

This week you will need to spend about half an hour every morning and every evening getting to know your rabbits (teddies!). Practice how you would approach and pick them up. Remember to stay quiet, and move gently and calmly. Look online and see if you can find videos of how to handle rabbits. You should be able to groom them, hold them safely and learn how to gently examine them to make sure they are well.

Most of the time should be used to just stroke and groom them to help with bonding. To start with, you might find it best just to spend some time quietly near your rabbits so that they get used to your smell and the look of you. Offering a tasty piece of fresh veg out of their ration when you handle them will help your rabbits see you as a real goodie rather than a possible predator! Try to look out for signs of when your rabbits have had enough.

All animals vary and your rabbits might not be interested in cuddles and grooming from you. If they are like this, remember to respect their needs and be happy just to watch them having fun together.

As well as the nice contact time, once or twice a day (depending on the time of year) you will also need to examine them for all the things we said in Chapter Seven that you need to look out for. You will need to briefly look at their bottoms and their underneath parts for poo or sores, but try not to take too long to do this because rabbits don't feel safe on their backs. You want the majority of the time they are with you to be a pleasant experience for both of you! You'll almost certainly find it easier to get a grown-up to help you look underneath. If one person holds the rabbit they can gently tip it upwards, always supporting its back legs, while you have a careful look round the back end. When you've got your real rabbits (if you decide to go ahead), this will be the time when you'll really start to get to know how your rabbits feel and how they behave. By handling them every day you'll soon spot when things are not right and if they feel skinnier or fatter than they should.

ARE RABBITS THE RIGHT PET FOR ME AND MY FAMILY?

Here is a table to fill in for this week. Once you are in the swing of things you'll find that most of the body check becomes part of your normal handling time and you add the bottom check as an extra. Oh yes, and don't forget to wash your hands afterwards!

Job/Day	Monday	Tuesday	Wednesday	Thursday	Friday	Saturday	Sunday
Time nearby, stroking, grooming AM							
Time nearby, stroking, grooming PM							
Full body check AM or PM							
Bottom check (should be done twice daily in fly season)							
Front teeth and claw check (once a week)	x	x	x	x	x	x	

Week Three — Fed, Watered and Neat as a Pin!

This week you'll be doing the dirty work! As well as feeding and checking the water is fresh and clean, you'll be learning about cleaning up. It's a rare person who really enjoys cleaning anything, so you can be forgiven for not looking forward to this – but it is a huge part of keeping pets. There is no getting away from the fact that some of what goes in has to come out, and it is up to you to clear it up! Handling poo and sometimes wee can make you poorly, so you need to make sure you know all about hygiene. Always wear gloves to clean out your hutch and the run, and always wash your hands afterwards. Look back at Chapter Three if you need to remind yourself about how much of which foods to give.

Of course, you still don't have any rabbits, so this week is about setting aside the time you need as if you had to do these jobs. Why not get your mum or dad to give you a boring job to do that would take about the same amount of time? Clean the bathroom, empty all the bins or do the ironing. You'll get an idea of the more boring side of being a pet owner, and you'll get massive brownie points at the same time! Also, do remember that most rabbits will be kept outside, so lots of these jobs will need to be done in the semi-dark and cold. You can't be a fair-weather pet owner!

Checklist for this week

Job/Day	Monday	Tuesday	Wednesday	Thursday	Friday	Saturday	Sunday
Check/give food AM							
Check/give food PM							
Give fresh water and clean bowls AM							
Check/freshen water and clean bowls if necessary PM							
Remove soiled bedding/empty litter tray once daily and top up bedding							
Wash food bowls and water bottles once daily (at least)							
Clean out whole hutch and replace bedding (once a week)							
Check toys, chews and enrichment and replace if necessary (once a week)	x	x	x	x	x	x	

ARE RABBITS THE RIGHT PET FOR ME AND MY FAMILY?

Safe	Poisonous

This is also the time to look online and find out about which plants are safe and which are poisonous for rabbits. Look at what's in your garden and identify the plants if you're going to let your rabbits roam or feed them plants from your garden. Here is a place to make your lists.

Week Four — EVERYTHING!!

And now for the grand finale. This week you will need to find a couple of hours every day in your hectic schedule to devote to your new pets. Fill in the rather large table below and add in your ongoing costs at the bottom. Most of all try to enjoy it because if you do get some rabbits you're going to be doing this for years, not weeks, and possibly even until you leave home!

Job/Day	Monday	Tuesday	Wednesday	Thursday	Friday	Saturday	Sunday
Time nearby, stroking, grooming AM							
Time nearby, stroking, grooming PM							
Full body check AM or PM							
Bottom check (should be done twice daily in fly season)							
Front teeth and claw check (once a week)							
Check/give food AM							
Check/give food PM							
Give fresh water AM							
Check/freshen water PM							
Remove soiled bedding/empty litter tray once daily and top up bedding							
Wash bowls once daily (at least)							
Clean out whole hutch and replace bedding (once a week)							
Check toys, chews and enrichment and replace if necessary (once a week)							
Costs £							

Time for the Family Debate

Over the last few weeks, and having read the whole of this book, you should now have some idea of what keeping rabbits is actually about. If you're like most people you'll probably be quite shocked. It's very rare for people to realise just how much time and money is needed to look after pets well. Lots of people can look after pets badly but I hope that now you will most definitely not be one of them!

You've probably been talking to your family about things as you've gone along, but if not, now is the time to do that. You can call a meeting and present your facts, like all the best detectives do. Because now you really do have everything you need to answer that question — and to answer it honestly. Things you might want to talk about at your family meeting:

- If you are under 16 years old, someone else in your family will be legally obliged to provide all these things for your rabbits, and they need to agree to that!

- Do rabbits, from what you've learned, sound like the sort of pets you'd enjoy keeping? If you thought they were something different, don't be ashamed to change your mind. That's the whole point of finding out all about them – to make the right choices.

- Can your family afford the costs you've found out about? Lots of people get embarrassed when talking about money, but now is not the time to be shy. If you can't afford them, don't get them.

- Did you have the time, energy and room to provide for all the things your imaginary pets needed? And if so, could you do that for up to ten years? If you're over the age of about 8, you may well be moving on before your rabbits die, so your family will need to carry on where you leave off. Are they willing to that?

- Is the whole family on board with the idea?

I hope that after all your hard work, you finally get the answer you wanted — but what about if you didn't? Time to ask the next question: what if the answer is no?

Chapter 9

As we said all the way back in Chapter One, you should never ask yourself what sort of pet you want; you should ask yourself what sort of pet you can care for properly. The fact-finding you've done up to now will hopefully have helped you work out if rabbits are animals you can keep healthy and, just as importantly, happy. As I said at the end of the last chapter, there is absolutely no shame in finding out that the answer is no. That is the point of your mission and the book: to help you and your family make the right, responsible choice. Not only will you have happy pets, but hopefully you'll have pets which make you happy too. Very often pets end up being given away because they were bought on an impulse with no research. This is very common in the case of rabbits. If kept alone, never handled and not neutered, they can be very unfriendly and unhappy animals and hurt a lot of children. This is through no fault of their own, but is simply because they are misunderstood and poorly cared-for. This becomes a vicious circle; the children don't want to touch them, and so they get neglected or given away.

So, if you have done your sums and learnt your facts and decided that rabbits are not the right pet for you or your family, then that is just as worthwhile as deciding to go ahead and buy some. You deserve a massive well-done, either way, and should be very proud of yourself. If you found that rabbits didn't tick your boxes or you couldn't tick theirs, it doesn't necessarily mean you can't have a pet — we just need to look at some alternatives, depending on what you were struggling with.

The charity I mentioned before (the PDSA) has come up with a great way to think about having pets and that is to think PETS: Place, Exercise, Time and Spend. Going through these four things for whichever animal you are thinking about is a good way to decide if you can keep them properly. On the PDSA website there is a great tool to help people find the right pet for their own situation, so do have a look at that as well. For now, we'll go one step at a time through PETS and see what pets might suit you best! Remember that this is just a pointer. You will still need to thoroughly research any pet you are thinking of buying. Even if a particular animal may seem appealing because it needs less room or is cheaper to keep, there may be other things about it that might put you or your family off.

Chapter 9

Place

Rabbits need way more room than many people expect. We are still too used to picturing the forgotten little hutch at the bottom of the garden. You may have been very surprised to see just how much they need, especially if you fancy keeping a group of them. So here are some pet ideas which might be better for you:

Guinea pigs.

Guinea pigs are lovely creatures and can make great pets. They need slightly less room than rabbits but two guinea pigs (they're social too) still need a hutch with at least 80 cm x 130 cm of floor space. As with rabbits, many guinea pigs are still kept in cages and hutches which are too small. They will also need an extra exercise area attached to their cage so if you didn't have room for rabbits you probably won't have room for these either. I'll leave that to you to decide!

Other 'small furries'.

Small furries are things like rats, mice, hamsters, and gerbils. Some of these certainly need less space than rabbits and would definitely be worth considering. There are some more exotic small furries kept as pets these days (such as chinchillas, degus and chipmunks) but some of these need pretty huge cages to let them express all their behavioural needs, so be very careful to do your research before you decide.

Cats and dogs.

These may be very different to what you were originally considering and in some ways need lots more space than rabbits, but they are still worth considering. You may have been expecting to keep your rabbits outside and found your garden just wasn't big enough. Because cats and dogs live in the house and just go outside for fun and exercise they might actually be better for you. Both cats and dogs have very complex needs, so make sure you can provide for those too.

Fish.

Fish are very calming, beautiful animals to watch and are very popular with lots of people. A fish tank will be much smaller than a rabbit hutch-and-run combination, but if you look into keeping fish, always consider how much room they would like because a tiny bowl can be just as bad as a little hutch. Please, if you do look into keeping fish, find out about where they come from. Some will be taken from the wild and this could be very damaging to the place they come from.

Exercise

In the case of rabbits, exercise is really included in the space requirements we've talked about, because you need to give them the space and enrichment to exercise when they want to. I like to think of this 'E' as energy instead. It takes a lot of energy to care for pets properly. That energy could well be walking a dog or playing with a cat but it could also be the energy you need to keep your rabbits' run new and interesting. It could be the energy you need to drag yourself into a cold garden to handle, groom and play with your rabbits every day. If, during your 'virtual month', that all seemed like a lot of hassle, what pet could you consider instead?

All pets need some commitment from you, but some need less interaction or different interaction which might be easier. Guinea pigs will need all the same handling and care as rabbits when it comes to energy expenditure, so are probably best avoided. Dogs in general needs lots of energy devoted to them and their exercise so if you're a bit on the lazy side, steer clear of them too!

Small furries.

Some small furries like hamsters are happy not to be fussed and might quite like being left alone. Most animals will get used to being handled but not all of them really need it and might prefer not to have it if given the choice. In fact a lot of the small rodents are happy to be left alone as long as their other needs are met, like being alone or with friends and being able to behave normally. You'll still need to expend some energy looking after them, but we'll look at that next.

Cats.

Cats are like teenagers; they're difficult to understand, they don't communicate well, they spend a lot of time sleeping, and they are likely to lash out at you for no apparent reason! Depending on the type of cat you have, you'll often find that cats like to interact strictly on their own terms and will come to you when they feel like having some affection. In this way they might need less from you than rabbits.

Fish.

Fish might be a good alternative, but make sure you know what sort of fish you are taking on and how much you will need to do to keep the environment right for them.

Chapter 9

Time

In your virtual month, especially in the last week, you should have found that your rabbits need a good one or two hours of your time every day and sometimes more. This might not sound a lot to many people but when you actually have to do it day in and day out it can quickly become difficult to find the time. You probably won't be surprised to hear that lots of non-pet-detective owners don't realise the time needed until it's too late, and pretty soon we're back to that neglected and forgotten hutch in the bottom of the garden or an abandoned rabbit.

Time is very precious to lots of people. We live in busy times. Lots of parents work, lots of kids do a million after-school activities and weekends disappear in the blink of an eye. Trying to find a spare two hours every day can be a massive headache for any family, even if you share the work. All pets need some time commitment from you and it's definitely worth considering right now if you found time an issue at all in your virtual month. It's better not to have a pet than to have an unhappy, badly looked-after one.

Time and energy go hand-in-hand. If you struggled to find the time for rabbits then dogs and guinea pigs will be the same. As we said with the small furries and cats, in some ways they may need less time. It will vary depending on the animal, and, as always, research will be important. For example, if you have a long-haired cat you might need hours to try and keep its coat under control but if you end up with an independent, short-haired moggie, some of them pretty much look after themselves. All the small furries will need some time for cleaning and feeding and again that will vary depending on the animal and the size of their cage, how they live and how dirty they are! Fish could well be another good alternative, but as with the energy required, some (such as marine fish) might need more time than others to keep their tank exactly right.

Time is a precious commodity, and so is money ...

Spend

The cost of keeping pets is probably the most massively underestimated thing of all when it comes to owners. Stuff like bedding, toys, and cages are easy things to work out and think about but people always forget things like vets' bills, vaccinations and, quite shockingly, the cost of food. Feeding an animal for two to twenty years can make a big hole in your wallet!

The PDSA PAW report in 2013 asked lots and lots of owners how much they thought it would cost to look after a dog, cat or rabbit for the whole of its life. Most people thought it would cost £1000-£5000 for a dog when in fact, depending on the breed it can be up to a staggering £31,000! Most people gave the same answer of £1000-£5000 for a cat, but the average is actually a whopping £17,000. I've already told you that the average rabbit costs £9000, but when owners were asked what they thought it would be, nearly all of them said about £1000. You can see why people are shocked when they actually get the animals! Money is another big reason pets end up being given away. As I said, some people are shy when it comes to talking about money, but if you're thinking of getting a pet it's absolutely vital you find out how much it is likely to cost and make sure your family can afford it.

From these numbers you'll have already guessed that if rabbits were too much of a strain money-wise, then cats and dogs are going to be a lot more of a strain.

Guinea pigs.
Guinea pigs need slightly smaller places to live and don't need any vaccinations. They tend to be less problem-prone than rabbits too, and live slightly shorter lives — up to around five years of age. All this added together means they can work out as a cheaper alternative to rabbits (usually!).

Small furries.
All the things that apply to Guinea pigs also apply to most of the small furries. In general, the smaller the animal, the shorter the lifespan. For instance, a rabbit can live up to ten years, a guinea pig around five years, a hamster two years, and a mouse up to a year. Also generally speaking, the smaller the animal, the smaller the cage you need — so on average smaller pets should be cheaper than rabbits to keep. Some of the more exotic ones will be very different, though, so do your research.

Fish.
Fish will be really variable depending on how many and what type you have. They have hugely different needs for things like water temperature, food and care. A couple of coldwater, freshwater fish may well be cheaper than two rabbits but a state-of-the-art tropical underwater heaven may not!

What if You're Worried You Can't Manage Any Pet?

Above all, please believe me when I say that it is better not to get a pet at all than to neglect one. Being a responsible pet owner is all about making the right choice and sometimes that means not getting one at all. Try not to be too downhearted. Look at all your options and also think about ways things might change with time. When I was very little we had no spare money at all but as we got older my mum and dad worked hard and trained and got different jobs and as time went by we found we could get a dog. Believe me, I pestered for a LONG time before that actually happened!

If you have friends who have pets, ask if you can spend time with their animals and help them with the jobs. You never know, they might be a bit bored of it and might love to have a helper. Just being around animals is a brilliant feeling so you might need to just take small opportunities when they come along.

Talk to your mum and dad about fostering pets. There are so many unwanted pets that lots of adoption centres often need people to foster animals while they are waiting for a home. This might mean having rabbits, guinea pigs, small furries and even dogs and cats if you can manage them, but just for a short time, and often with help from the charity with the costs. You'll get some animal time but also be doing a really good deed too.

You could also find out about charities like Hearing Dogs for Deaf People. Some of the centres ask people to care for a dog in the evenings and over the weekends while it's being trained. This means if your mum and dad are at work during the day the dog just comes to you for the times you're all at home. They have lots of different volunteering options on their website so you might find something that is perfect. A temporary arrangement might be a great compromise for you and your family.

So here we are, almost at the end of your journey into the wonderful world of rabbits and being an A-grade owner. All that's left to say is ...

Chapter 10

By the time you get to this part of the book, you will have worked hard and learned exactly what it takes to be an excellent pet owner. Let's think about all the amazing things you have done and achieved since the first page:

- You've learnt about how animals can make us happy and what it means to be a responsible pet owner, including the serious stuff about the law!
- You've found out how wild rabbits live, what makes them happy and what keeps them healthy.
- You've found out what food is best for rabbits, which plants are safe and which are poisonous, how much food rabbits need, and how to tell if rabbits are too fat or too thin.
- You've discovered just how much room rabbits need in order to live happily and what sort of environment will keep them safe. A HUTCH IS NOT ENOUGH!
- You've learnt that rabbits hate to be alone and need the company of other rabbits to feel safe, secure and surrounded by friends.
- You've found out that it's very important to let animals behave normally. Rabbits need to be able to play, hop, run, jump, binky, dig, stand up tall, chew, groom, graze, explore and of course, eat their own poo!
- You've learnt about vaccinations, obesity, tooth disease, fly strike, myxomatosis, RHD and all the things you need to look out for to spot a poorly rabbit before it gets too bad. You've also learnt that rabbits, because they are prey animals, don't always show pain like other animals, so it's up to you to know what's normal for them.
- You've done extra research, spoken to vets and nurses, been to pet shops, looked online, done lots of maths and maybe even made your whole family sit down together to discuss this whole pet-owning business.
- Hopefully you've even gone the whole hog and made yourself feel a bit silly by wandering round the garden pretending to look after two or more pretend rabbits.

Chapter 10

THAT is a very impressive list of things and that is why you should feel very proud of yourselves. I am ecstatic that you bought or borrowed this book and read it and I am very proud of you. You may feel like kids are ignored sometimes and you might feel sometimes like no-one really listens to your opinions, but I'm going to let you into a secret: you kids can change the world. Let's face it, grown-ups have messed up pet-keeping for hundreds of years. They think they're too busy to do research and lots of them think they know everything already!

Just imagine if you told your mum or dad all the things you've learned about how to actually care for rabbits. I bet they would be astounded. I bet you could teach them some things. They might think it's fine to keep them on their own or with a guinea pig! But now YOU know better. You might not feel like you can change the world but if all the children in the world learnt what you have and really took it on board, then pet-keeping would transform overnight. Old, wrinkly vets like me would smile and put our feet up — our workload would halve in an instant because of all those well cared-for pets.

I've told you repeatedly that you should never ask yourself what sort of pet you want, but now we're at the end of your journey I think it's time for you to do just that. You see it's very important that you ask yourself what sort of animal you can care for properly but you do also need to consider what you're looking for in a pet because you and your family need to be happy too. You might have discovered that you can perfectly care for a hundred rabbits but if you wanted a pet that would cuddle up on your lap every night, a hundred rabbits might never make *you* happy! Owning pets is a team game so make sure you all talk about your options. If rabbits aren't right for you, have a look at the other books in the series and keep learning. All animals are fascinating even if you don't end up with one.

If you've decided rabbits would make you happy and that you can keep them happy and healthy, hang onto the book. Unless you've got a brain the size of a planet you might want to remind yourself of things when you get your new bundles of joy. If rabbits now seem like the worst choice in the world, then pass the book on. You could give it to a friend or sell it for some pocket money. If you've got a friend with a rabbit you don't think is being very well looked after, you could slip the book into their school bag as a nudge in the right direction. If that friend gets their rabbit a companion as a result, you've taken one more step towards changing the world.

At the start of the book we said that living with animals can be wonderful. I hope the books in this series will help to guide you and your family to what will be a fantastic friendship and a time that you will look back on with big smiles and a mountain of happy memories.

All that remains for me to do is to tell you again how brilliant you are and to award you with your Detective Certificate — proof that you now know pretty much everything there is to know about the needs of rabbits.

WELL DONE!

THIS IS TO CERTIFY THAT

- -

HAS LEARNT PRETTY MUCH EVERYTHING
THERE IS TO KNOW ABOUT CARING FOR RABBITS
AND BEING AN EXCELLENT PET OWNER